Work 245

鸟巢

A Bird's Nest

Gunter Pauli

[比] 冈特·鲍利 著

[哥伦] 凯瑟琳娜·巴赫 绘

朱 溪 译

上海远东出版社

丛书编委会

主　任：贾　峰

副主任：何家振　闫世东　郑立明

委　员：李原原　祝真旭　牛玲娟　梁雅丽　任泽林

　　　　王　岢　陈　卫　郑循如　吴建民　彭　勇

　　　　王梦雨　戴　虹　靳增江　孟　蝶　崔晓晓

特别感谢以下热心人士对童书工作的支持：

匡志强　方　芳　宋小华　解　东　厉　云　李　婧

刘　丹　熊彩虹　罗淑怡　旷　婉　杨　荣　刘学振

何圣霖　王必斗　潘林平　熊志强　廖清州　谭燕宁

王　征　白　纯　张林霞　寿颖慧　罗　佳　傅　俊

胡海朋　白永喆　韦小宏　李　杰　欧　亮

目录

Contents

一只北美红雀正开始筑巢，为下蛋做准备。一只知更鸟看到北美红雀付出诸多辛苦，问道：

　　"你是觉得只要把树枝堆在一起，把雏鸟放在里面，鸟巢自然就能支撑住吗？"

A cardinal starts building its nest, and is getting ready to lay eggs. A robin, observing the tremendous effort the cardinal makes, asks,

"Do you think that just by shoving sticks together, your nest will hold once you have growing nestlings in there?"

一只北美红雀正开始筑巢……

A cardinal starts building its nest ...

... twigs, grasses and strips of bark ...

"我想用这些细枝、草和树皮条制成一个杯子一样的巢。它可以支撑我的鸟宝宝的体重，并且对宝宝的皮肤来说很柔软，直至鸟宝宝顺利长出羽毛。"

"我无法想象，你的巢不用泥土就能支撑好几个星期。"

"What I want to create with these twigs, grasses and strips of bark, is a cup. One that will hold the weight of my chicks, and that will be soft for their skin until they grow feathers."
"I can't imagine your nest holding together for weeks without some mud."

"嗯，它的确可以！我先用喙压碎细枝，等到可以扎成一束后，我再用爪子把它们做成理想的杯子形状。"

"我不需要有关筑巢的教导。我们一生要筑二三十个巢，而且每个都几乎一模一样。泥土就像砂浆，把一切都搅在一起。"

"Well, it does! I first crush the twigs with my beak, and when these can be plied I will use my claws to push them into the desired cup shape."

"I don't need any instructions on building a nest. We build twenty or thirty over a lifetime, and they all look exactly the same. Mud is like mortar, it holds everything together, in one place."

我先压碎细枝……

I first crush the twigs ...

······等待一场大雨······

... wait for a good rain shower ...

"对我们北美红雀来说，一个鸟巢只能用一季。我们每年都会在有食物，并且方便照顾幼鸟的地方筑新的巢。"

"我们知更鸟是等待一场大雨，然后用喙叼着泥浆放到树上。你知道吗，我们建造巢需要飞一百多趟呢？"

"For us cardinals, a nest lasts only one season. We build new ones every year, close to where our food is, and where we can keep an eye on our chicks."

"We robins wait for a good rain shower, then take beaks full of mud up into a tree. Did you know that it takes me more than a hundred flights to build my nest?"

"这么说，你们巢的核心是泥土。而我的巢外部有硬的细枝，然后是叶子，里面还填塞了树皮，以及一层草和松针。"

"我把所有细枝编织在一起，一旦固定到位，就用泥浆将它们粘在一起。然后我用肚子将巢形成完美的摇篮，再用我的毛发作为衬里给蛋保温。"

"So, the core of your nest is mud. Mine has strong twigs on the outside, then leaves, and the inside is lined with bark, and then a layer of grasses and pine needles."

"I weave all the twigs together and, once in place, I cement them with mud. Then I use my tummy to shape my nest into the perfect cradle, before lining it with hair to keep my eggs warm."

······我的有硬的细枝······

... mine has strong twigs ...

下雨的时候用泥筑的巢会怎么样呢？

What happens to a mud nest when it rains?

"你知道怎么给蛋保温？太有才啦！"

"还不止这样，我们还能建造非常坚固的鸟巢。我们的蛋也许很轻，但我们一只雏鸟的重量可以是整个鸟巢的四倍。"

"可是下雨的时候用泥筑的巢会怎么样呢？"

"You know how to keep eggs warm? That's ingenious!"

"That's not all, we also build very strong nests. Our eggs may weigh little but our nestlings can each weigh four times more than the whole nest."

"But what happens to a nest made of mud when it rains?"

"只有一种选择——我必须展开翅膀坐在我的巢穴和雏鸟之上，这样孩子和巢穴才能保持干燥。"

"这可不简单，"北美红雀回答道，"我的巢就简单多了。"

"不过，对我而言最糟糕的是那些不速之客。"

"你是说那些捕食者吗？"

"There is only one option – I have to sit on my nest and nestlings with my wings spread out, so the kids and the nest remain dry."

"That cannot be easy," Cardinal responds. "My nest is much simpler."

"The worst for me, though, are all those unwanted visitors."

"Predators?"

展开翅膀，让孩子保持干燥

With my winds spread out, keeping kids dry

引来了苍蝇、虫子和虱子……

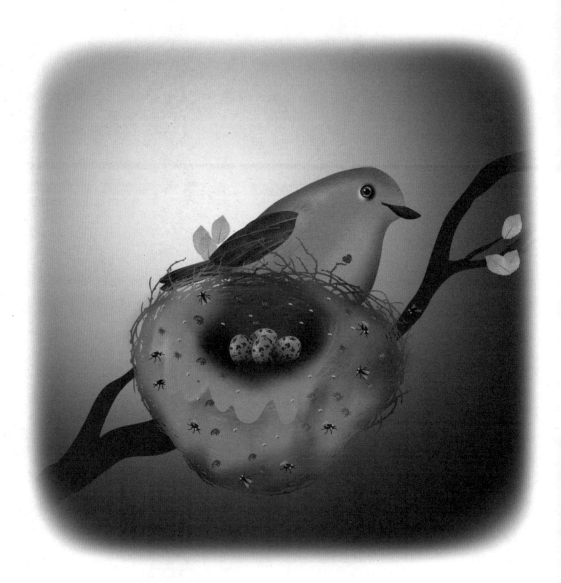

Attracting flies, and mites, and lice ...

"哦，我们倒是可以应对它们的。问题是泥浆封住了巢的地板，所以便便就堆积起来，引来了苍蝇、虫子和虱子……"

"哦，不！这些虫子以周围的废物为食，并且无休止地繁殖，你的鸟巢对鸟宝宝来说肯定会变得不安全的。"

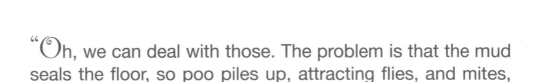

"Oh, we can deal with those. The problem is that the mud seals the floor, so poo piles up, attracting flies, and mites, and lice…"

"Oh no! With all those insects feeding off all the waste around, and multiplying all the time, your nest must become unsafe for your chicks."

"那时就得继续前进了。当我的妻子开始建造新的鸟巢时，我会把小鸟赶走，逼他们学会飞行，这样我们才可以养育下一批宝宝，重新开始。"

　　……这仅仅是开始！……

"And that's when it is time to move on. I kick out my chicks, forcing them to learn to fly, while my wife starts building a new nest, so we can start all over with the next batch of babies."

... AND IT HAS ONLY JUST BEGUN!...

...... 这仅仅是开始！ ......

... AND IT HAS ONLY JUST BEGUN! ...

The creativity of birds' nest building is unparalleled in Nature. Cupped nests are the most common, while others adhere to buildings, or are constructed as platforms, dug into the earth, or are scrape nests.

鸟儿筑巢的创造力在大自然中是无与伦比的。杯状巢最常见,其他形状的鸟巢有的附着在建筑物上,有的构造成平台嵌入地下,还有的是浅坑巢。

Cowbirds lay their eggs in other birds' nests, and screech owls use cavities hollowed out by other birds or animals. Pigeons breed like rabbits, nesting up to eight times a year.

牛鹂在其他鸟的巢中产卵,而角鸮会利用其他鸟或动物挖的洞。鸽子像兔子一样繁殖,每年筑巢八次。

Bald eagle chicks will stay in the nest for up to 100 days. Killdeer chicks hatch fully feathered, and will leave the nest after only 5 minutes. Birds raised in captivity will build nests without ever having seen one.

白头海雕雏鸟最多会在巢中待 100 天。双领鸻幼鸟孵化出来就长满羽毛，仅 5 分钟后就会离开巢穴。被圈养的鸟即便从来没见过鸟巢也会筑巢。

Birds nesting near human habitations will use man-made objects such as paper, string, nails, pieces of wire, and fabric for nesting material.

在人类居住区附近筑巢的鸟类会使用纸、线、钉子、金属丝和织物之类的人造物作为筑巢材料。

Some birds line their nests with specific plant material, selected for its ability to inhibit mites and other parasites, while others will find and place pieces of shed snake skin in or near their nests to deter predators.

有些鸟在巢中会用特定的植物材料做内衬，用来抑制螨虫及其他寄生虫，而另一些鸟则会找蛇蜕放置在巢中或附近以威慑捕食者。

Hummingbirds build tiny nests and line them with spider webs. Bald eagles build huge nests, used year after year. These can be 2 to 3 metres in diameter, and a metre or 2 deep, weighing up to 2 tons.

蜂鸟建筑微小的巢，并用蜘蛛网做内衬。白头海雕修筑巨大的巢穴，年复一年地使用。它们的直径可达2到3米，深1到2米，重达2吨。

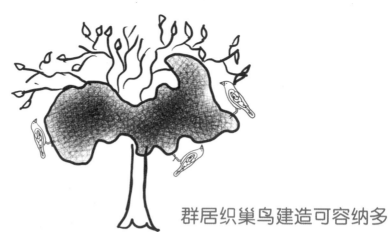

Sociable weavers build communal nest complexes that can house up to 400 families, and that can last for a 100 years, housing many generations. These have thatched roofs, for insulation against heat and cold.

群居织巢鸟建造可容纳多达 400 户家庭的公共巢穴复合体，而且可以持续使用 100 年，容纳许多代子孙。这些巢穴设有茅草盖的屋顶，用于隔热和防寒。

Mallee fowl build the biggest nests in the world: 3 metres high and 10 metres across. The bird creates a compost heap and when it is warm enough, lays its eggs on top, and monitors incubation.

眼斑冢雉能建造世界上最大的鸟巢：高 3 米，宽 10 米。眼斑冢雉会形成积肥堆，当肥堆足够温暖时，便把卵产在顶上，并监控孵化情况。

Does the idea of moving house with every new baby appeal to you?

每生一个宝宝就搬一次家，这个主意吸引你吗？

How about transporting construction material in your mouth?

用你的嘴巴来"运输"建筑材料感觉如何？

Would you kick your offspring out of the home you've built for them?

你愿意把自己的后代赶出你为他们建造的家吗？

How long would one be able to live in a house without a toilet?

在一个没有卫生间的房子里，人可以住多久呢？

Do It Yourself!

自己动手!

How many different types of nests have you seen in your life? Can you differentiate between them? Let's ask our friends and family members to join us in walking around city gardens, parks and in forests to find different types of nests. Take a look at towers, chimneys, bridges, barns, and abandoned buildings. Photograph as many different nests as you can find and make a list of them. Now compare your lists. Discuss which nests you think are best protected against wind and rain, and why. Keep on adding to your lists, and before you know it, you may have a combined total of a hundred different types of nests.

到目前为止，你见过多少种不同的鸟巢? 你能辨别它们吗? 和亲朋好友一起去城市的花园、公园和森林里寻找各式各样的鸟巢吧。看看塔、烟囱、桥梁、谷仓和废弃的建筑物。把各种鸟巢拍下来，并做个清单进行比较。讨论你认为最能抵挡风雨的巢穴及其原因。添加列表，不知不觉之中，你也许能积累到一百种不同的鸟巢资料。

学科知识
Academic Knowledge

生物学	北美红雀吃种子和水果，但喂幼鸟吃昆虫；求偶喂养，雄性喂养伴侣；北美红雀是雀形目鸟；北美红雀会秃头；北美红雀是肉食性的；雄鸟和雌鸟唱歌；对鸟巢的研究称为鸟巢学。
化 学	砂浆是由泥和黏土制成的；石膏灰浆，称为熟石膏；硅酸盐水泥的原料包括石灰和沙。
物 理	巢体现了几何形状、弹性和摩擦力的相互作用；巢是一个不对称的弹簧：被推时变硬，释放时变软；迟滞现象，系统的表现是随之前所做的操作变化而改变；筑巢过程中使用保温材料。
工程学	由雌鸟独自建造的杯状巢由枯叶和苔藓制成，内衬毛发；北京奥林匹克体育场的形状像一个鸟巢，需要42 000吨钢材来支撑这种随机结构；不对称弹簧，推压时坚硬，释放时柔软；北美红雀将楔形巢筑成叉子形状；使建筑结构不透水。
经济学	像北美红雀这样的鸟类可以消除甲壳虫、蟋蟀、蝈蝈儿、蚂蚱、叶蝉、苍蝇、蜈蚣、蜘蛛、蝴蝶和飞蛾等昆虫；有能力只使用当地材料建造房屋的经济学。
伦理学	北美红雀和知更鸟作为鸣鸟被圈养；它们是一夫一妻制的鸟类。
历 史	砂浆首先在公元前10000年在杰里科用于建筑，在公元前8000年在达列赫用于建筑；在公元前2500年的埃及，石灰石块被泥浆和黏土或黏土和沙子的砂浆固定。
地 理	知更鸟在英国是最受喜欢的鸟；北美红雀原产于北美和南美；北美红雀是伊利诺伊州、印第安纳州、肯塔基州、北卡罗来纳州、俄亥俄州、弗吉尼亚州和西弗吉尼亚州的州鸟。
数 学	纵横比，直径除以巢的长度得出其最大（或稳态），密度；体积密度，单位为KN/m^3。
生活方式	建立一个无法传给下一代的房屋；不要让房屋里有会造成严重健康问题的小害虫。
社会学	知更鸟会在夜晚的路灯下唱歌；庇护孩子，直到他们被"踢出"家，开始学习过独立的生活。
心理学	北美红雀会攻击镜子中的自己；依赖性强的巨婴；让子女保持依赖的父母。
系统论	使用后可重新融入大自然的建筑物设计；从可再生资源中选择建筑材料。

情感智慧
Emotional Intelligence

知更鸟

知更鸟欣赏北美红雀的筑巢技能，但对鸟巢的强度存有疑问。她坚信泥土用作砂浆是最好的。知更鸟告诉北美红雀更多细节，强调自己会努力筑起坚固和舒适的鸟巢，比如为了筑巢，她可以飞一百次。她直截了当地批评了北美红雀。当北美红雀对她的筑巢技术提出疑问时，知更鸟会提供实用的信息。她承认的确是存在问题，雨水和糟糕的卫生条件会导致虫害。再一次，她的回答非常直接，自信地表明了对害虫问题的解决方案：将小鸟赶出去并开始筑新的巢。

北美红雀

北美红雀心平气和地接受了知更鸟的批评。他首先解释了自己正努力实现的目标。然后，他详细介绍了自己筑巢过程中的细节，并清楚地阐明他优先考虑鸟巢要离食物较近。当知更鸟自夸多次将泥土运到巢中时，他以务实的方式做出回应，没有给出评判。他表现出对知更鸟的巢和雏鸟的关心，当知更鸟担心雨水、废物和害虫时，他表现出了同情心。

艺术
The Arts

筑巢是一门艺术。以团队合作的方式，用树枝、树叶或泥土，试试能否用手做出鸟儿用喙和爪筑成的鸟巢。仔细研究实际巢穴的大小、形状和构造方式，然后开始建造。找一些苔藓、树叶、头发、羊毛和毛皮来装饰你的鸟巢。这不仅考验你的工程和建筑技能，也是对鸟类为幼鸟建造安全温暖家园的致敬。

思维拓展
Systems: Making the Connections

鸟巢就像育婴室，是为鸟蛋和雏鸟提供庇护的临时场所。大自然赋予鸟类创造各种最佳筑巢方案的自由。有大量的筑巢技巧及巢穴形状让人印象深刻，但其原则却是不变的：鸟类会使用当地可用的资源，从环境中获取的任何东西最终都会返回大自然。鸟巢不仅适应当地气候和生态系统，而且巢的形状和结构也与环境和谐，甚至到了鸟巢难以被人发现的程度。

筑巢是一种与生俱来、代代相传的技能。筑巢显示出了某种内在智慧。混合使用的材料非常出色，因为鸟巢不仅要考虑其内容物的重量和不断增加的尺寸，还要画龙点睛地使之又舒适又温暖。鸟类仅依靠喙、爪、翅膀和腹部就能创造出所需的形状和空间。这为我们重新思考现有资源，适应当地条件，并确保所有物品都可以组合与拆解提供了灵感。

鸟类超凡的筑巢本领，启发科学家使用未经筛选的材料来满足可持续发展的需求。现代建筑工程领域与鸟类的智慧相去甚远。鸟类在利用城市环境方面表现出高度的适应性，在所有受威胁物种中具有最好的成活率。鸟类确保后代需求与安全的能力及其物种生存的方法，的确值得我们所有人尊重。

动手能力
Capacity to Implement

邀请一些朋友和你一起到附近找一些黏土和泥，并用它们来筑巢。每个巢要足够大，能方便地容纳三到四个鸟蛋。只允许用手来混合黏土并为巢塑形。可以加一些水，让黏土更有延展性。仔细规划巢的形状，同时牢记雏鸟在其中的安全性和舒适性。关键是要用你在自然界中发现，并且能回归大自然的材料。一旦对巢的构造和内衬满意了，让黏土慢慢干燥。在巢上放置重物来测试其强度。如果支撑得住，再洒一些水，看看巢会不会分解。相互比较所做的巢的强度和耐用性。

故事灵感来自
This Fable Is Inspired by

卡罗拉·迪里奇
Karola Dierichs

卡罗拉·迪里奇曾就读于德国不伦瑞克工业大学、瑞士苏黎世联邦理工学院，2009 年以优异的成绩毕业于伦敦建筑联盟学院的新兴技术和设计专业。2020年她获得德国斯图加特大学计算设计与建筑研究所工程学博士学位。如今，她是柏林魏森湖艺术与设计学院的材料编码学教授，也是柏林洪堡大学卓越集群项目"活动着的物质（MoA）"的成员。在此之前，她曾是卓越集群项目"建筑一体化计算设计与建造"下设的计算设计与建造研究所（ICD）的研究助理。她创造的星状颗粒置入适当位置，可以形成类似鸟巢的结构。

图书在版编目（CIP）数据

冈特生态童书.第七辑：全36册：汉英对照 /
（比）冈特·鲍利著；（哥伦）凯瑟琳娜·巴赫绘；
何家振等译.—上海：上海远东出版社，2020
ISBN 978-7-5476-1671-0

Ⅰ.①冈… Ⅱ.①冈… ②凯… ③何… Ⅲ.①生态
环境－环境保护－儿童读物—汉英 Ⅳ.①X171.1-49

中国版本图书馆CIP数据核字（2020）第236911号

策　划	张　蓉	
责任编辑	程云琦	
封面设计	魏　来　李　廉	

冈特生态童书

鸟巢

[比]冈特·鲍利　著
[哥伦]凯瑟琳娜·巴赫　绘

朱　溪　译

记得要和身边的小朋友分享环保知识哦！
八喜冰淇淋祝你成为环保小使者！